D0906774

ILLINOIS PRAIRIE DPL

IT'S SNOWING!

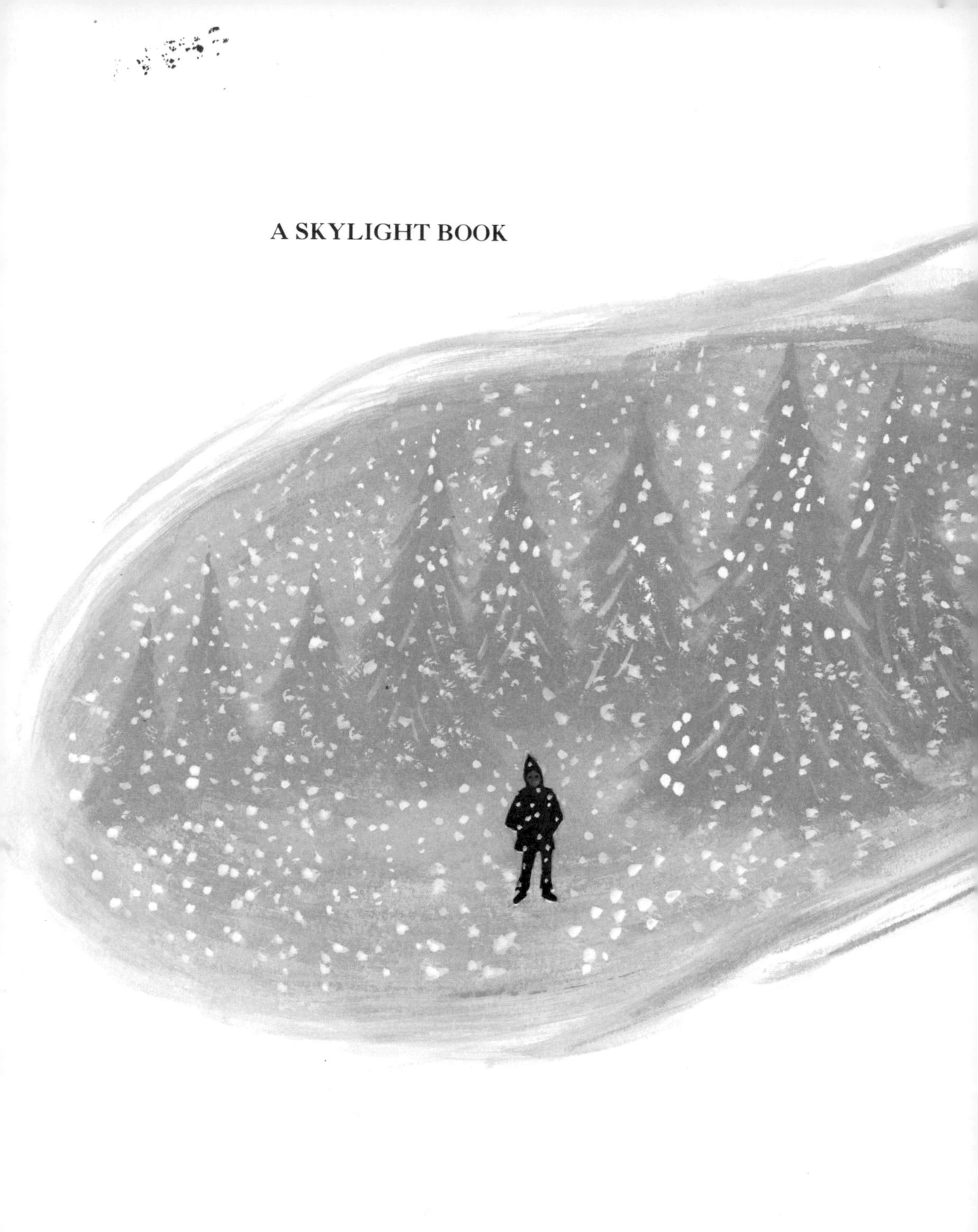
A SKYLIGHT BOOK

IT'S SNOWING!

Margaret Cosgrove

DODD, MEAD & COMPANY
NEW YORK

To my good mother and father, as always

Printed in the United States of America

1 2 3 4 5 6 7 8 9 10

Library of Congress Cataloging in Publication Data

Cosgrove, Margaret.
Its snowing!

Includes index.
SUMMARY: Discusses the qualities of snow, how it is made, and how it can be helpful or dangerous for people, animals, and plants.
1. Snow—Juvenile literature. [1. Snow]
I. Title.
QC929.S7C74 551.57'84 80–14254
ISBN 0–396–07851–6

Contents

Look, It's Snowing!

The day is cold and dark. There is a feeling that something is going to happen.

Tiny white "flowers" begin to drift down through the air—one, two, then many. Faster they tumble, and faster still. Soon the world looks as if some sky-giant had dropped an enormous cloak of white that covers the cars, the buildings, the ground. It has "sleeves" for all the chimneys, "skirts" for the roofs to wear, a "sweater" for every car. Streetlight poles, fences, and fire hydrants are all dressed in white. There are many-fingered "gloves" to fit

each tree. And all the white garments are formed without sewing or seams, with no need for zippers or buttons. It is a cloak of snow worn by the whole countryside.

Snow is the cold fluff that sifts into cracks and holes, and tries to get in the tops of your boots. It can tickle your face, or can sting. It levels off street curbs, and smooths down steps into a slippery slide. It makes fire escapes look like stairways that belong in a palace. It creates whirlwinds in corners that dance and spin, and die down—then rise up like ghosts and sink again. The snow whirls and curls; it makes hollows and hills.

Sometimes snow blows straight across fields for miles and miles, out where the land is flat and open. Snow fences that make shadows like combs are put up to keep it from drifting across highways and country roads. Curtains of snow blow so thick and dense that the horizon—even buildings and trees—seems lost forever.

A winter snowstorm enfolds the world in a garment of white. It holds some of the heat of ground and buildings beneath it, and is truly a cloak of warmth.

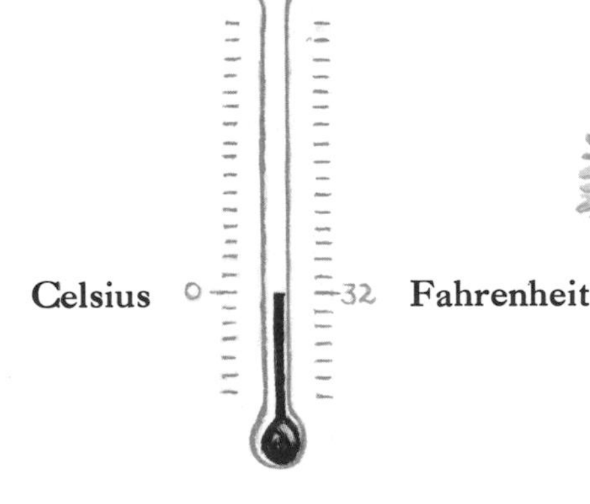

How Is Snow Made?

It is nearly impossible to find two snowflakes, out of all the billions, that are exactly alike. Study one on your mitten, or on a black cloth. Using a magnifying glass is a good way to help you see a snowflake.

Snowflakes are formed when it is cold enough for the water vapor in clouds to freeze. A cloud is made of water vapor, which is like fog or mist. You are inside a cloud when the world around you is foggy. Once in a while snow is formed in very moist, cold air near ground level, instead of coming down from a cloud.

Each crystal of snow needs something to start growing on. This is usually a tiny speck of dust or ice. Thousands

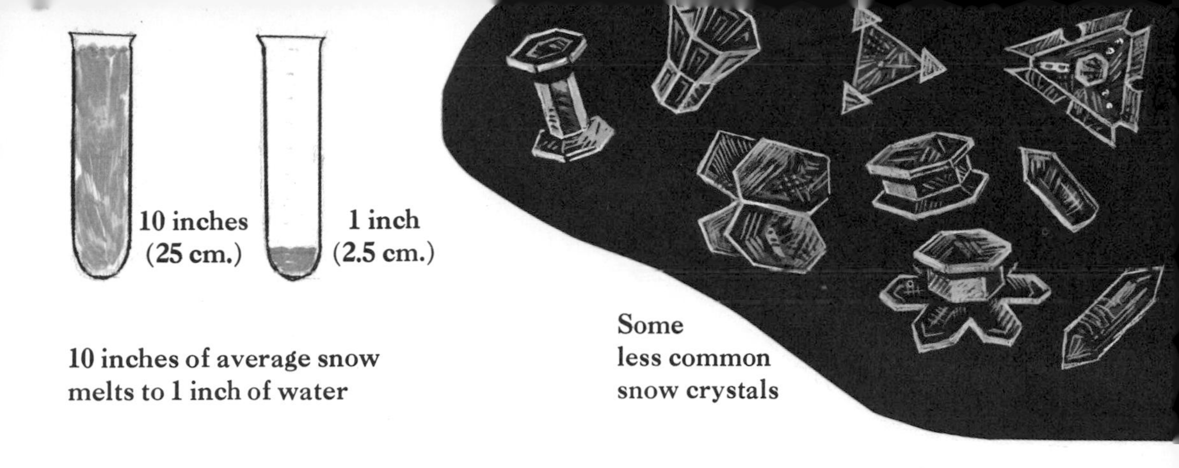

10 inches of average snow melts to 1 inch of water

Some less common snow crystals

of water vapor droplets attach themselves to the speck as they freeze, lining up in a certain way to form a single snow crystal. Crystals often stick together as they fall to earth and form bigger snowflakes—when the temperature is right. Some have been seen as large as the palm of your hand!

Many snow crystals are needle-shaped, branchy, or like little bars or bullets. The form they grow into depends on the height and temperature of their vapor clouds, how much moisture is in and around them, how fast they

Common snow jewels

Wilson Bentley photographing a snow crystal

fall, and the kind of weather they pass through on their way to earth.

For hundreds of years people have wondered why so many snow crystals are six-sided. In 1880, a Vermont farm boy was given a microscope for his fifteenth birthday. From then on, Wilson Bentley spent much of his life, during the cold New England winters, working out ways to photograph snow crystals through his treasured instrument. He captured the beauty of thousands of snowflakes on film. They can still be seen and admired in his book called *Snow Crystals*.

A scientific study of snowflakes is difficult, since most are formed high above the earth, and may change while falling. They melt and break so easily that scientists have tried to study their formation in laboratories. In the 1930s Ukichiro Nakaya began investigating snow crystals. A university in Japan helped him build the first cold chamber, a small room that could be chilled down to very low temperatures. After many experiments, Nakaya found that he was better able to grow snow crystals on tiny dots of rabbits' hair than on anything else. Now, in many countries, much study continues, especially about fallen snow and the changes the crystals go through packed together deeply on the ground.

Nakaya with snowflake-making device in his cold chamber

Snow Is Many Colors

Snow is the whitest of whites—but not always.

Actually, snowflakes are clear as glass, but the way they reflect light makes them appear white. And there are three ways snow can be black:

One is when it gets dirty.

One is when you look at it against the sky, when it looks like dark little propellers whirling.

One is when tiny plants, so small they must be seen through a microscope, grow in fallen snow and turn it black. Such plant life, called algae, can also change snow's color to red, yellow, blue, and even green and pink! There are more than a hundred kinds of these algae known to occur in snow in various places in the world.

In the shadows snow appears violet or blue. When the sunrise glows on it, it can seem pure gold, and the sunset can turn it to rose. In the night, snow seems to be deepest blue or purple. Sometimes it looks like diamonds with silvery twinkles glittering over it—like handfuls of stars tossed down.

Listen to the Snow

A heavy snowfall muffles noises—the shouting of children having fun in it, the screech of sirens, the honking of horns. Everything sounds as if your ears were stuffed with cotton.

Sometimes falling snow sings and zings.

It can sound hushy, slushy, and squooshy.

It can whistle and hiss as it blows.

Snow can squeak and creak. It can crackle and crunch under your boots. Sometimes it sounds like lots of little feet tap-dancing on the windows.

Or it can be perfectly silent.

Listen to the silence. See if the snow will whisper its secrets in your ear.

Snow Falls from Pole to Pole

A small snowstorm may drop its white cover only on a few fields or a little town as it travels over. But another snow cloud system could be so large it would darken, then whiten, a whole state or even several of them as it sweeps on its way. It may move onward slowly, or perhaps swiftly, until its water vapor has been used up or the weather on high has changed.

Around the center of the earth, near the equator, there is rarely any snow. But even here there are countries that have it on high places. Kilimanjaro, a mountain in East

Africa, is one of these. Its snow-capped peak can be seen on a hot day from the countryside below. Some mountains in warm countries like Algeria and Greece also have snow on their peaks.

The areas of the world between the frigid polar regions and the warm lands near the equator are called the *temperate zones*. Here is where snow falls in winter, while the summers are warm.

But it is at the South and North Poles where we picture the most snow. The South Pole is on the continent of Antarctica, cold beyond belief, and almost completely covered

Mount Kilimanjaro

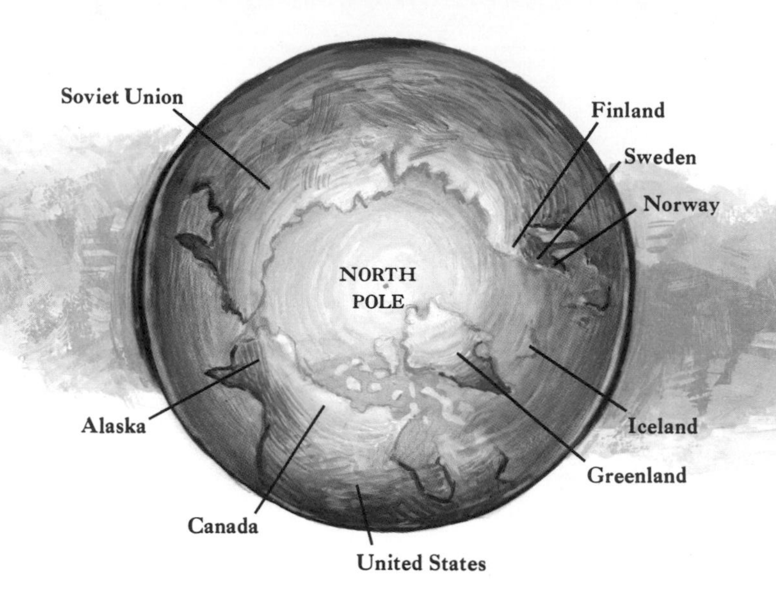

by glaciers of ice. The North Pole is quite a different sort of place. There is only the deep Arctic Ocean here, spotted by some islands and filled with drifting ice packs. But, surprisingly, not much snow falls at either place, because extremely cold air cannot hold much moisture from which snow crystals form. Yet these regions are very snowy, because so little snow ever melts. It adds up year after year, century after century, pressing down under its own weight to become ice.

The South Pole is uninhabited by human beings, except for a few scientists and explorers. But people have learned to live in the lands ringing the North Pole—in

Alaska, Greenland, and the Soviet Union, and also in Norway, Sweden, and Finland. They recognize a great many kinds and uses of snow, and have many names for it.

There are different words for snow lying on the ground, for falling snow, wind-packed snow, snow where reindeer graze, snow reindeer lie down in, snow for igloos, and dry, powdery snow. Eskimos, especially, have a large vocabulary for it. All snow is called *annui*. Windblown snow is *upsik*, snow caught in trees is *qali*. Fluffy snow is *theh-ni-zee*. Nothing is more important in the lives of these Northern peoples than *annui*.

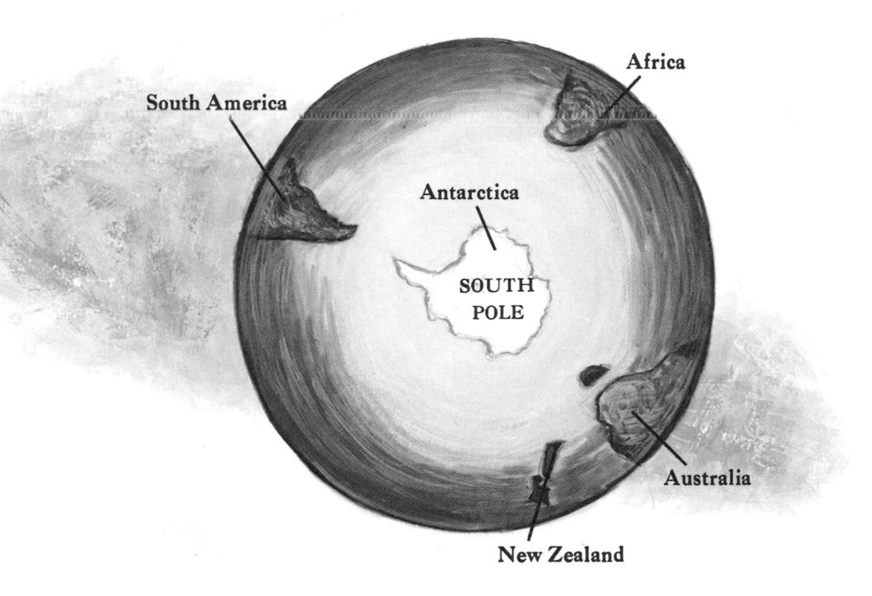

Snowflakes Have Great Power

What happens when billions of snow crystals mass together? When alone, they are weak and frail. But heaped on tree branches, they can become heavy enough to weigh them down, and even break them off.

In mountainous countries thousands of tons of snow can break away from high slopes and crash downward with a deafening roar. Whole villages have been lost under one of these *avalanches*. Houses are battered as if they were made of match sticks, trains are swept away like toys, and herds of animals have perished helplessly when gigantic avalanches thundered down on them. Human beings

have helped cause avalanches in some places by cutting down trees and forests that would have held back the snow.

People sometimes become buried under an avalanche too. Search teams today include "avalanche dogs," trained to run about over the snow as they try to pick up the scent of a person lost somewhere beneath. Saint Bernard dogs were the first kind used. They began going with monks in the high, steep Alps of Europe well over 500

An avalanche

years ago to help travelers through the dangerous mountain passes. Other kinds of dogs are also trained nowadays. People can sometimes be saved after hours of being buried in snow.

A snow blanket spread over hills and fields can look peaceful and quiet—until spring arrives. In March or April, heavy snows can pour off more of their meltings in a sudden thaw than waterways can handle. Torrents of water come roaring between the banks of brook or river, overflowing onto the land. The result is floods that sometimes wash away houses, cars, trees, animals, people.

But the whole countryside is often helped by melting snow in gentler ways. Snow warmed by spring sun sinks into the ground, supplying young plants and new crops until they are stronger. Hundreds of trickles and rivulets join into streams, plunge over waterfalls, and fill rivers to keep them flowing. Without this water from melted snow a great many fish and plants would die.

When snow falls faster than it melts away—over many centuries—its own weight packs it down until it becomes huge, solid masses of ice, deeper than the very tallest buildings. These giant ice cakes, called *glaciers*, look like gigantic frozen rivers, waterfalls, or fields. There are thousands of miles of these glaciers in cold places of the world, such as Greenland and Antarctica. And they once covered a large part of what is now the United States. Locked into each glacier are millions of tons of frozen water that was once many delicate, dancing, little snow crystals.

A glacier

Snow Can Be Dangerous and Difficult

It is hard for driver and passengers to decide what to do when their car becomes stuck in heavy snowfall. They must either remain trapped in it and hope for help to come, or abandon it and try to trudge to a safer place.

Deep snows can make serious problems for trains, and blizzards cause trouble for boats. Sometimes plane passengers have a strange experience when their pilot is able to climb up through clouds to fly high above a storm. They can look down on snow clouds from blue and sunny skies. But other airplane passengers may be stranded in airports for days at a time when a snowstorm takes over their world.

Snow and ice make their trouble, but the many tons of salt thrown on them each year to help melt them do their share of damage also. Salt dissolved in the water from melted snow cracks up road surfaces. It kills plants and even many large trees when it washes to their roots, and some salt remains in the soil to keep causing problems. When it seeps through the ground into streams and ponds it can cause certain algae to grow all out of bounds and spoil the water quality. Birds and smaller animals have been poisoned by drinking salty snow water. Sand thrown on snowy, icy streets to keep tires from slipping is often blown by the winds, and clogs up street drains.

Walkers, unable to see what lies under snowdrifts, may step in a hole, or stumble and be hurt. They may lose their way in a blizzard. Farmers have become lost between house and barn. Fingers and toes may suffer frostbite when they become too cold. Warming them slowly and gently is the best treatment. Do *not* rub these parts with snow, as is often advised.

Sunny brightness on all that white can cause "snow blindness." Eyes become dazzled and tired from looking at too much of it. This goes away after the person has rested his eyes for some time indoors. Eskimos made "sunglasses" of bone, with long narrow slits in them that kept out most of the light.

Out in the country, people are sometimes unable to enter or leave their houses when heaps of snow keep doors from opening. Or an avalanche from a roof may coast down on one's head when the door does open! Digging a path to barn or car, or going somewhere for food, may be nearly impossible. So can getting to work, to hospitals, or to fight fires. Ropes are sometimes strung up between houses and barns, or along sidewalks in some windy, snowy cities, for strugglers in blizzards to cling to.

Cold Snow Spreads a Warm Blanket

Snow puts funny caps on weeds, tall hats on berries, and pillows on fence posts. It also spreads a real blanket, one that keeps the ground beneath it often many degrees warmer than the air above. It does this by trapping the earth's warmth in the spaces among fallen flakes. Without this, many smaller plants would die, torn by the wind, frozen in the icy air. A snow blanket also keeps the ground temperature less changeable, and it is the freezing and thawing, freezing and thawing of bare ground that causes the death of many plants and animals living in and on it. Without snow's protection, soil would be ripped from roots and blown away.

Some plants have a ring of leaves around the bottom of

Ruffed grouse

their stems, called *leaf rosettes*. These can stay alive and green through frosty weather because they hug the ground, snug under their white quilt. Many plants even melt snow around them when they push up through it, long before spring has arrived. Snowdrops are very early flowers that can be seen in the snow.

When seeds are scattered on crusty snow the birds can find them and pick them up, but new or blowing snow quickly hides such food. Birds leave their footprints on soft snow like mysterious signs, or alphabet letters, stamped all around. A thirsty bird can drink a little beakful of snow when no water can be found.

Bluejay

Cardinal

A rosette under snow **Snowdrops**

A bird feeding station with seeds, suet, and crumbs placed in it helps birds through the hard times. Birds that would chase each other away in spring and summer eat from the same lunch counter in winter. When certain ones appear, such as the big, flashy bluejay, smaller ones flit away. But they will be back! When you see the beautiful red cardinal, look to see if its mate is nearby. Many male and female birds stay near each other all winter long.

A few large birds squirm right down into the snow to keep warm and protected. The grouse makes a flying dive into soft snow—like leaping into a fluffy feather bed. It plows itself in deeper with the thick leg feathers it grows for winter.

If you see dark patches on the snow that look as if someone had spilled pepper there, investigate them closely—with your trusty magnifying glass if possible. You will be in for a surprise! Hundreds of tiny "snow fleas," or springtails, come up through the snow in places on sunny days, and travel back down through it to the ground again at night. Their tails fold under their bodies, making them hop like little jumping jacks when released.

Small worms looking like short snips of brown or black thread can also sometimes be found twisting about on the snow. These, like the springtails, feed on tiny plants living there. These are algae, similar to the kinds that give color to snow in some places, and are too small to be seen very well without a microscope. Small spiders dine on the springtails and worms when they can find them, and birds like to find the spiders. Snow can be a lively place!

Springtails live in snow, even in the Arctic and Antarctic

Snow Time Brings the Hungry Time

To larger wild animals, snow can be helpful or harmful. The slender legs of deer, and the longer legs of elk and moose, are like stilts holding their bodies above deep snow. But most hoofed animals—including bison and wild mountain sheep and goats—have feet too small and pointed to walk easily in snow. They sink down helplessly in drifts, or when needing to run through a deep white cover. Only the reindeer and caribou of the Far North have bigger hooves, which spread apart for easier walking in snow. These animals wander far and wide seeking soft snow to paw through for the kinds of plants they eat. Each adult must try to smell out about twenty-five pounds of plants each day to fill its empty stomach.

Though some bears slumber in caves or beneath toppled old trees in winter, others bed down in deep, open snow. Steam has been seen to rise from their big, warm bodies as snowflakes melted around them. Some animals—especially otters—cannot resist playing in snow. They seem to love going for long slides down slippery hillsides!

But to most other animals winter brings problems too

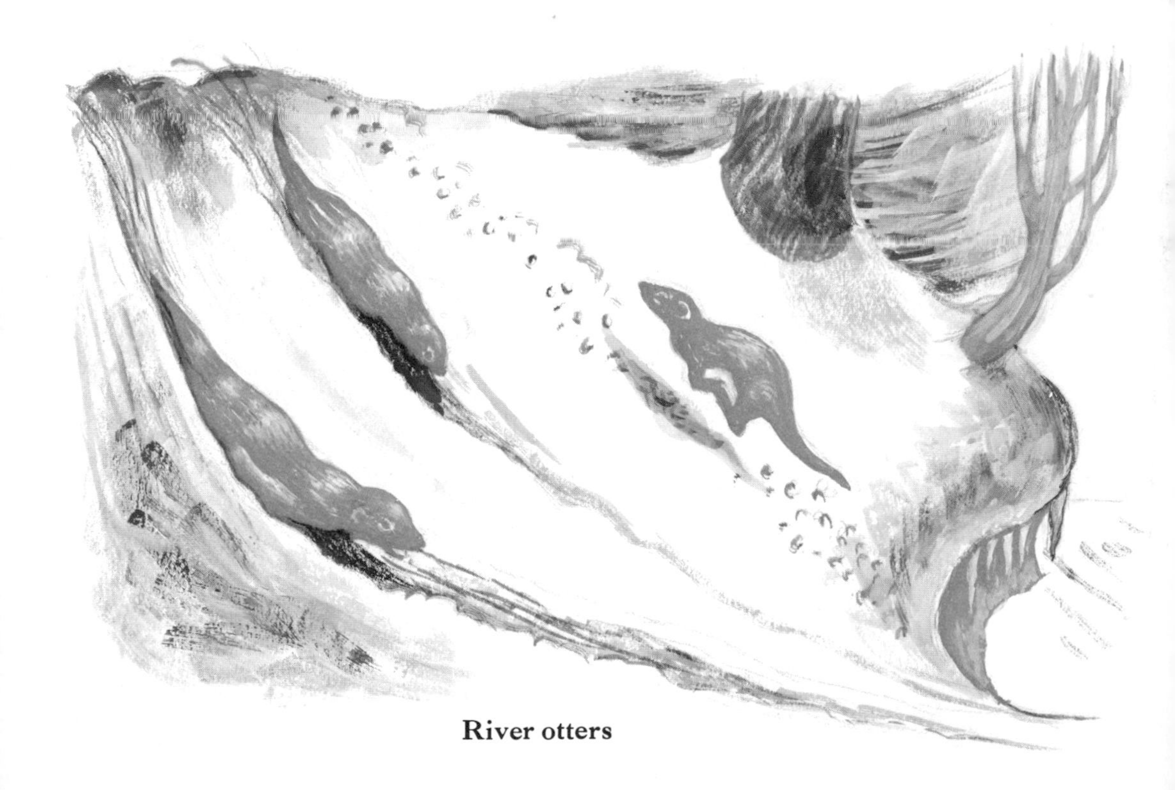

River otters

Lynx

serious for fun. Deer, elk, and moose often gather in groups, and their many feet trample the snow in paths that make walking easier. Where a network of paths is made, it is called a *yard*. But food—bark, twigs, and some evergreen branches—is used up extra-fast by so many together. It is a time of hardship.

Winter is the season when wolves once played an important and necessary part in nature. They took the weaker animals that would have had the hardest time surviving the cold and snow. Wolves did not attack the strong—or steal the farmer's sheep and calves. They preferred to go after the weakened animals that wandered alone, far from man's domain.

The lynx of the North is the only wild cat good at winter living. Its heavy coat makes it look bigger than it is, and its huge, furry paws are almost as good as snowshoes. But still it is no match for the snowshoe hare (a cousin to the little brown cottontail rabbit) when it comes to running in snow. A lynx has to take leaps through the snow. It sees a hare—it crouches and freezes in its tracks like a statue. The hare comes closer to take a look—did it see something, or not? The lynx makes a powerful bound for it.

All hunting animals make many tries before catching a meal. The lynx sometimes waits hours, hardly moving, for a hare to pass by. Life is hard in the winter.

The cottontail stays brown all year

The snowshoe hare turns from brown to white in winter

A Hidden World Under the Blanket

Atop the snow, the world often looks like a silent sleeper. But what is all this going on beneath the blanket?

Snow protects many small lives from hungry eyes. Muskrats make dark passageways beneath the snow from the creekbank of their home to where they know plants still grow green and tasty. Even smaller creatures build tunnel systems under the white cover. Remember that it is warmer down here, because this blanket holds in the earth's heat.

Some of these creatures under the snow are shrews and voles. They are mouse-sized and common enough, but seldom seen by our eyes, even in summer, because of their secretive ways. They build whole road systems under the

Foxes eat many berries and plants

snow, as do meadow mice. Grassy nests and little storehouse rooms stocked with seeds and nuts are here and there. The forever-hungry shrew seeks them out.

The velvety moles spend much of their lives underground, but even they—especially the star-nosed mole—occasionally come upstairs to join other little animals under the snow. Many of these small-world citizens build their own hidden hallways, but they also scuttle through those made by others. What must it be like to bump into another scurrying little body here, between walls that are white but always dark?

Up above their snow roof may be foxes, coyotes, or even wolves on the prowl. But these animals are valuable in keeping down the numbers of the little creatures forever having babies. The light-footed fox, especially, can run on crusted snow. He stops and listens. He awaits

faint squeaks or squeals from below, then sniffs for a warm, mousy smell. A quick jump on hind feet, breaking through the ceiling of some surprised tunnel-dweller, and he may have a meal. Sometimes the fox is the winner, sometimes his little prey gets away.

Foxes and owls watch for small air vents that appear at night in the snow like little chimneys. The air below gets stale and a vole needs a few breaths of fresh air once in a

while. They may be his last, if a sharp-eyed owl or fox is on the lookout.

Weasels go about things differently. They leap into snow and paddle through it like swimmers, their slim bodies twisting through someone else's runways until they find their dinner.

And then there is the truly sleeping world in rotting logs, beneath stones and dead leaves, down even in the ground. Of all the snails, turtles, snakes, queen bees, worms and grubs, beetles and other insects, toads and salamanders hidden here, far more survive the winter when snow spreads a thick coverlet over the world than when the ground lies bare.

When winter comes, fur and feathers, shells and scales help keep in animals' heat. Scarves, coats, and mittens keep in yours. The snow holds in the earth's warmth.

Please Do

—have lots of fun with snow! Who can resist making a snowball?

—make yourself into a snow angel. Lie on your back in the snow and carefully brush your arms up and down to make wings. Then swing your legs back and forth over the snow to form the gown.

—shovel a driveway or sidewalk for someone, or help them dig out their car. Or go to the store for someone who can't get out, or walk their dog.

—treat yourself to snow ice cream. Mix just a little warm honey or maple syrup in a bowl with very clean snow, and eat it with a spoon.

—play a jolly game of Fox and Geese. Make a big circle in the snow, and then make three paths across it so it looks like a wheel with six spokes. Then play tag on it, staying just on the paths. You're safe when you're standing in the middle, where no one can tag you.

—ride on a sled, coaster, or toboggan. You can also ski or snowshoe, even trying to make your own skis or

snowshoes from thin boards or slats from crates if you have none.

—make a snowman. Or build a snow fort.

Snowshoe

Please Don't

—throw hard snowballs, or aim snowballs at heads, car or bus drivers, or windshields.

—get lost in a snowstorm or blizzard.

—track snow into the house.

—leave things (sleds or ice skates) lying around where snow can cover them up, or throw away trash into the snow.

—let ice or salt freeze on your dog's paws, or get stuck between its toes, or cut its feet. (Some Eskimos made booties for their sled dogs.)

—stay out too long when fingers and toes have begun to get cold. Or fling off your coat outdoors just because you have gotten too warm from playing or shoveling.

—step on ice on a pond or stream unless you are absolutely, positively sure it is safe.

—go at shoveling too hard or too fast. Stop and rest now and then. (Adults unused to this kind of hard work often find it brings on a heart attack.)

Most of all, don't forget to have a good time in the snow!

Index